Confessions of a Catholic Scientist

*The Quest for a Malaria Vaccine
and the Quest for God*

A Journey of Discovery

Robert Theophilus

Published by

Leonine Publishers LLC
Phoenix, Arizona, USA

ISBN-13: 978-1-942190-36-3
Library of Congress Control Number: 2017954613

Printed in the United States of America
10 9 8 7 6 5 4 3 2 1

Visit us online at www.leoninepublishers.com
For more information: info@leoninepublishers.com

Table of Contents

PART I

In With a Bang

"Isn't it a miracle?"

By this the attending physician did not mean a miracle in the usual sense that the birth of any child is in some way miraculous. Rather, he meant that it was a miracle that I was alive, inasmuch as my mother had conceived twins and at three months into the pregnancy my twin sister died. Unexpectedly and without anyone's knowledge, my mother carried a dead baby for six months without a miscarriage. And so, on July 16, 1945, the Feast of Our Lady of Mount Carmel, I left the safety of my mother and entered the outer world. On the same day and at about the same hour that I was born and approximately 1000 miles southwest of where I was born, Robert Oppenheimer and his crew exploded the world's first atomic bomb, and so you might say that I came in with a "bang." Shortly thereafter, the world called upon the great Catholic philosopher, Jacques Maritain, to help draft the United Nations' Universal Declaration of Human Rights.

My father grew up in Winnipeg's tough "north end." His mother died when he was nine years old, and when he was sixteen he had to leave home. In 1939 he joined the Canadian Army but a bad foot eventually led to a medical discharge. He then worked for what was at that time one of Winnipeg's better hotels, although it has since been converted into a hostel for homeless men.

Although Manitoba is located in the center of Canada, bordering on North Dakota, it was settled long before many of the regions in Eastern Canada. Explorers searching for the Northwest passage to Asia often ended up in Hudson's Bay on the north shore of the future province. The first Europeans to reside in Manitoba were fur trappers employed by either the Company of Adventurers of England Trading out of Hudson's Bay, or their archrival, the North West Company, headquartered in Montreal. This changed when the Industrial Revolution took hold in Europe and wealthy land owners in Scotland and Ireland suddenly found it more profitable to evict their tenant farmers, in order to raise sheep to provide wool for the English mills. One of these wealthy land owners, the Earl of Selkirk, had compassion on the displaced farmers and, using his own resources, helped to resettle them in North America; notably in Nova Scotia and in Manitoba's Red River Valley. Near the end of the nineteenth century, farmers also came from the steppe regions of Europe, including many from the Ukraine, and this is where my mother's story begins.

My mother was born in 1922 on a pioneer farm in the region between Lake Winnipeg and Lake Manitoba near the town of Fraserwood. My grandfather had a quarter section of land that has been described by the renowned

Ukrainian-Canadian historian, Michael Ewanchuk, as consisting of "spruce, swamp and stone." My mother had nine brothers and sisters and, along with parents and a grandparent, they lived thirteen people in a two-room house. There was no running water or electricity, and in winter when you had to relieve yourself you went outside at thirty degrees below zero to use a snow bank. Nevertheless, they were happy. My grandfather grew up on a farm and knew how to be self-reliant. When one of my mother's brothers split his lip wide open, my grandfather knew exactly what tree sap to use to bind the lip together so that it healed without a scar. At Christmas and Easter he could cook up a feast on a simple wood stove and, on the weekends, all of the neighbors came over to get a free haircut from "Waiko" Sam. When the war broke out, my mother's brother Johnny went to train with the Canadian air force at the nearby Gimli air base, and it is one of my great disappointments that he passed away before I could ask him whether he knew Thomas Merton's brother, John Paul Merton, who was training at Gimli at about the same time.

My mother required only six years to complete the eight grades taught in their one-room school house and took grade nine by correspondence course, which she passed with honors. Subsequently, she had to leave school and the farm for the city. It was the middle of the depression and the government was offering eight dollars a month to well-off families so they could hire domestic servants. So my mother worked as an indentured slave for the next three years. Eventually she was employed as a chamber maid in the hotel where my future father was working, and that is how they met.

PART II

Close Encounters

When I was about five years old I had a close encounter with death. My parents took me to one of the beaches along Lake Winnipeg and a neighbor told my mother not to worry, that he would look after of me. For some reason he allowed me to float on an inner tube and, when the water below me was over my head, I fell off the tube and went straight to the bottom. When I was at the point of starting to suck water into my lungs, someone reached down, grabbed me by the hair, and pulled me up. It was the neighbor, but, inasmuch as the water was opaque, how he knew exactly where I was located below the surface, I can only attribute to the grace of God.

Like most other youngsters, my childhood and early teenage years were spent focusing primarily on school work, as I was interested in it and, besides, my father told me, on a number of occasions, that if I did not get an A average he would kill me. With this incentive I usually did well enough in high school to be exempted from writing final exams. I still had a somewhat feral streak in me, however, and at the end of high school when I received an award before all of the parents, I thought it oh so *au courant* to subsequently go off with my friend to shoot pool in a billiards hall located in one of the most dangerous slum areas of the city.

After high school I had no idea what the university was all about other than it was where I needed to go to continue my education. After a very confusing first year, I decided that I liked chemistry enough to enroll in a very demanding three-year honors chemistry program. Right at the start, I had what can only be described as a second death-like experience. In the first session of the physical chemistry lab, old Dr. Campbell addressed the class and gave us a stern warning against breaking an expensive Beckmann thermometer.

During one of the subsequent labs, I had to return the Beckmann to the lab instructor, but he was occupied and told me to put it back in the rack that held the thermometers. When I tried to push the rack back into its cabinet, one of the thermometers became jammed against the top of the cabinet. Now I had spent the previous summer working at a bottling plant where I had to heave heavy drink cases all day. Well the rack that held the thermometers felt like a drink case and so my arm automatically shoved and, of course, broke the dreaded Beckmann. How well I remember Dr. Campbell as he came roaring into the lab demanding to know, in front of all my classmates, "Where's the man who broke the Beckmann?" Fortunately, he didn't throw me out and I was able, eventually, to make amends. At the end of the year he gave us a fairly difficult, long-answer physical chemistry exam and, I hope that I won't appear supercilious to say, that he, as well as all of my classmates, who, I am sure, continuously thought of me as the dolt who broke the Beckmann, were surprised to learn that I was the only one who scored 100% on his exam.

PART III

Hard Times

I focused on the physical and inorganic aspects of chemistry and became quite proficient at molecular orbital theory, quantum mechanics, and chemical physics. For my Master of Science degree I applied all of this to a study of nuclear magnetic resonance. Then I made what I consider to be one of the biggest mistakes of my life. I wanted to pursue a doctorate in chemistry that was in some way related to medical research. I also wanted to remain at the University of Manitoba as I was very attached to, and, therefore, did not want to leave, our family dog. There were only two options that were apparent to me: first, biochemistry, which I disliked greatly, and second, one of the chemistry professors was synthesizing anti-cancer drugs. Even though I was not at all competent in organic chemistry and disliked it almost as much as biochemistry, I chose the latter. (Such thoughtlessness is certainly food for thought.) A few months into the program I realized what a mistake I had made, but I foolishly said to myself that I have started this and so I am going to finish it. Well it almost finished me instead.

To make matters worse, we were quite poor and the room that I inhabited in my parent's house was a type of pantry off of the kitchen. It had space for a small closet, plus a desk and a single bed placed one against the other and nowhere else. I could either sit at the desk or lay on the bed. I occupied this pantry from age 18 to age 27. The last four years, age 23 to 27, were the hardest. In addition to gaining very little consolation from my work, many of my colleagues at the university were getting married, but I had no friends, and during the long winters could not even go out for a walk, as it was usually freezing cold (minus 30° F to minus 40° F), pitch black outside with the snow up to the rooftops.

My younger sister would often come to interrupt my solitary confinement and tell me that she was concerned about one of my cousins, who was the same age as I, as he was still single. She wondered whether there might be some way that she could help him to meet some young ladies. The fact that her brother was in the same boat did not seem to merit any consideration. One good thing that did occur during my four-year ordeal was that I was constantly being challenged in the laboratory by a Protestant who had memorized a number of verses from the King James Bible that appeared to support his position. This prompted me to start reading the Catholic Bible and other Catholic books and thus began, albeit in a rather anemic and inchoate fashion, the greatest and most intellectually satisfying adventure that I have ever known. At that time, however, as my Catholic faith was not strong, I became suicidal, and it is a wonder that I managed to hold out to the end.

Time passes, even if not *tempus fugit*, and eventually liberation day arrived. I had applied for and was awarded

a three-year post-doctoral fellowship from the Medical
Research Council of Canada and, given my aversion to
organic chemistry, I had arranged to do my first year of
post-doctoral research studying electron spin resonance at
the University of Toronto. While all of this was happen-
ing, the University of Manitoba had arranged for several
members of a university in Montreal to come to Winnipeg
to establish a Department of Immunology. Just before I left
for Toronto, the head of the proposed new immunology
group came to the Department of Chemistry and gave a
seminar on immunochemistry and I was very interested in
what he had to say.

When I arrived in Toronto I was happy to see that they
also had a fairly large immunology department. I attended
their seminars and became convinced that I had to find
some way to transition from chemistry into immunology.
This fire was further fueled by the fact that I was not very
impressed with the person that I had come to work with in
Toronto. He did not appear to know very much and all he
would do was lay back in a chair in his office with his eyes
closed. When asked what he was doing he would always say
that he was dreaming up new projects. I decided to leave
this dreamer who, not surprisingly, was fired about a year
later. During summer vacation in Winnipeg, I went to the
immunology head and asked whether he would accept me
into his department. Because I still had two years left on
my Fellowship he was agreeable to the idea but wanted me
in his own lab. So that fall I left Toronto to return to Winnipeg and thus began my forty-year sojourn in the exciting,
challenging, and constantly evolving field of immunology.

For those readers who are not familiar with this discipline, a brief explanation is in order. Immunology is the study of the body's immune system and how it protects the body against infection and disease. It also tries to elucidate the mechanisms that lead to various autoimmune disorders such as multiple sclerosis and diabetes and that prevent the immune system from eradicating various cancers. To accomplish this it characterizes the numerous sub-populations of white blood cells that are found not only in blood, but also in secondary lymphoid organs such as the spleen and lymph nodes. It seeks to establish the function of each unique cell type, how these cells regulate each other's activities, and whether they do this by direct cell contact or by the secretion of soluble mediators. It also examines the molecular and biological bases for self-nonself discrimination and for immunological memory, which makes human vaccination feasible.

PART IV

Allergic Reactions

When I first joined the department I did not know any immunology, so I started by auditing almost all of the graduate courses and by doing a prodigious amount of reading on my own. Fortunately, modern immunology was just starting to burgeon as a discipline and I got in on the ground floor. After a few years, I applied for and was awarded a five-year Medical Research Council Scholarship which brought with it an appointment as Assistant Professor as well as the opportunity to write my own research grant applications.

Almost the entire department was studying the immune responses that mediated the allergic reaction and so this was where I directed my efforts. Because allergy is a type of hypersensitivity, the goal was to study the sub-population of lymphocytes, referred to in those days as suppressor T (Ts) cells that could dampen the allergic response. I wanted to know whether the Ts cells recognized the same region of the protein allergens as the helper T (Th) cells that promoted the allergic response. To accomplish this, I used insulin as a model allergen because insulin was available from a number of distinct animal species and the slightly divergent amino acid sequences of these species' variants were all known. By comparing the cross-reactivity of the Th cell and Ts cell responses to the insulin molecules, I hoped to be able to map the regions recognized by these cells. (Of course, there are currently much more sophisticated methods to map an immune response.) The system had the additional advantage that insulin is also a major antigen targeted by the autoimmune response leading to diabetes and, therefore, characterizing immune responses to this antigen would have broader applications.

In the summer of 1980, I was asked to attend the International Congress of Immunology in Paris, France. This was exciting in that it not only offered an opportunity to meet many of the world's leading immunologists, but it would also be my first trip to Europe. I had a direct flight to London from Winnipeg on a big Lockheed TriStar. I shared a hotel room in Paris with Dan Conrad, who used to be a postdoctoral fellow in our department and was, at that time, an assistant professor at Virginia Commonwealth University. His real name was the German spelling of Konrad and, as far as I can recall, he was able to trace his roots in the United States all the way back to the Revolutionary War. In the evenings, we visited the more famous Paris landmarks, including the Eiffel Tower, the Arch of Triumph, and the Palace of Versailles.

Although I was not scheduled to give a presentation at the conference, at one of the sessions I was informed that the head of my department had requested that I be allowed to give a talk. I was told this about ten minutes before I was scheduled to speak, and so I had to quickly dig up some slides and scribble down a few notes. Given that I had never spoken previously at a major immunology conference, and that I was so unprepared, when my turn came and I stood before the audience I was shaking so hard it is a wonder that I didn't black out with my hastily prepared notes flying off in all directions. The material that I discussed was novel and interesting, however, so by the grace of God I managed to survive both the presentation and the ensuing questions, although I must have been quite a sight.

One evening all of the conference participants were invited to a rather elegant garden party on the grounds of the Palais du Luxembourg. We were treated to a string quartet playing classical music as well as unlimited quantities of French pastry, ice cream, and Champagne. One expected Le Roi Soleil to show up at any minute. On the weekend, I attended Holy Mass at Notre Dame Cathedral, where there are bishops entombed, starting from the fifteenth century, and this gave me a sense of the continuity of the people of God.

PART V

Buona Giornata Roma

After the conference, I decided to remain in Europe for my usual two-week vacation. I spent a few days with Dan Conrad in Switzerland, as his post-doc and his post-doc's wife both had family there who were eager to show us a good time. They regaled us from the Jura Mountains to the Alps, including an unbelievable cable car ride to the top of the Matterhorn. One of the parents had been diagnosed previously with terminal cancer and it was suggested that he buy a chalet in the Alps so that he could enjoy his last six months. He agreed and when Dan and I were there he had been enjoying his last six months for over twenty years. It is amazing what a little mountain air can do for you. It was at this juncture that I began to experience, for the first time, the consequences of Crohn's disease, although I did not realize it at the time. Subsequently, Dan and I parted company and he returned to Paris while I took a night train to Roma.

Of course the first order of business was to visit the Vatican and St. Peter's Basilica. Many people do not know that, although the Pope fulfills most of his liturgical functions at St. Peter's, his cathedral church as the Bishop of Rome is St. John Lateran. St. Peter's exceeded my expectations with regards to its size and artistic beauty—a true tribute to Michelangelo and the myriad of other great artisans who contributed to it. The magnificent high altar with its towering baldachin is located directly over a reliquary containing the bones of St. Peter.

In 1939 when some workmen were excavating the crypt below the main church, to create a space for the tomb of Pope Pius XI, one of the workmen fell through the floor to find that St. Peter's is located on top of a fairly large and ancient underground cemetery. The discovery of St. Peter's remains in the so-called graffiti wall is a fascinating archaeological detective story. I went down into the crypt and was able to pray at the tombs of St. Peter, Pope John XXIII, Pope Paul VI, and Pope John Paul I. I have read that after the death of Pope John Paul I, his secretary recounted that the Holy Father had told him that the cardinals who elected him had made a mistake and that he was going and the foreigner (non-Italian) was coming. When asked what foreigner, the Holy Father replied: the one who sat opposite me during the conclave. Subsequently, it was revealed that this was none other than Karol Wojtyla, the future Pope John Paul II.

Another item high on my agenda was a visit to the Sistine Chapel. The portrait of the creation on the high ceiling of the chapel consists of a series of frescos painted by the great Michelangelo. These frescos were made from watercolor on fresh plaster, so that when the paint penetrated the plaster and the plaster subsequently dried, the painting became a permanent part of the ceiling. One can imagine Michelangelo laying on his back, day after day, on a high scaffold painting furiously so that he could complete his oeuvre before the plaster dried, and yet he managed to produce such masterworks. The great artist also painted "The Last Judgment" on the far wall of the chapel, where God is shown completing what He began at the first moment of creation, separating light from darkness.

Of course I also had to see the Roman Colosseum. This ancient (circa AD 70-80) amphitheater, about the size of a modern-day football stadium, is where the Holy Father begins the Stations of the Cross on Good Friday. The interior consists of a series of concentric circles with many arches and passageways. One could easily picture the young St. Therese of Lisieux, the Little Flower of Jesus, along with her sister Celine, escaping from their tour guide to scramble down to the very bottom in order to venerate the stones made sacred by the blood of the early Christian martyrs.

One evening, just prior to leaving Roma to return to the Red River Valley, I came out of my hotel and walked right into a gigantic parade of Italian Reds. It stretched from sidewalk to sidewalk and as far down the street as I could see. Many of the marchers were shaking their fists in the air and chanting something, although the only word I could recognize was *fascista*. I quickly headed in the other direction only to run into two truckloads of Italian police and two truckloads of Italian soldiers with machine guns. I can tell you that I was very frightened but fortunately it was soon over and I was able to return to the hotel. Two days later I was back on Alitalia airline heading home.

PART VI

Great Discoveries

It was about this time that two key discoveries reported in the scientific literature were to revolutionize every aspect of medical science, including vaccine development. The first was the elaboration of a method to produce large quantities of monoclonal antibodies. Normally, immunization of an animal will induce the activation of several different clones of B cells that will differentiate to become antibody producing cells. Consequently, the response and the ensuing antibodies are polyclonal in nature. If the B cells target several different sites on the immunogen, this leads to even greater diversity. To generate monoclonal antibodies, Kohler and Milstein used polyethylene glycol to fuse polyclonal antibody producing murine B cells with murine cancer cells (myeloma cells). This procedure yielded hybrid cells (hybridomas) that retained the properties of both the B cells (antibody production) and the myeloma cells (uninhibited cell growth). The hybrid cells were then selected for HAT Medium, as neither unfused B cells nor unfused myeloma cells survive under these conditions. The hybridomas were subsequently cultured at limiting dilution where a maximum of only one cell is placed in each of the wells of a 96-well culture plate. Because only one cell (one clone) was plated per well, the antibodies produced under these conditions were monoclonal. Each positive clone was then grown up to large numbers in culture flasks to yield large amounts of pure monoclonal antibody.

Monoclonal antibodies have myriad uses. For example, within a tumor, there are molecules on cells of the immune system and on the tumor cells themselves that can inhibit the immune response to the tumor. Blocking of these inhibitory molecules with specific monoclonal antibodies can liberate the body's immune system so that it can eradicate the tumor. Similarly, monoclonal antibody against the inflammatory mediator TNFα can ameliorate autoimmune diseases, such as rheumatoid arthritis, Crohn's disease, and ulcerative colitis.

The second momentous discovery was the development of a technique for cloning a gene. The method is quite complex, but, to summarize very briefly, DNA containing the required gene is isolated and then cut into fragments using a restriction enzyme. These fragments are ligated into DNA plasmids (small rings of DNA) which are next inserted into bacteria and the bacteria containing the gene of interest are cloned. The gene is next transferred into an expression vector where it can be replicated and then transcribed into RNA and subsequently translated into protein. Scaling up this procedure facilitates the production of large amounts of pure protein, which can then be used, for example, to immunize recipients against a pathogen expressing the protein. How this technique was employed to generate a vaccine against malaria will be described later.

PART VII

B sheba and T sheba

One of the post-docs attending my graduate immuno-biology lectures was a medical doctor from mainland China. During my conversations with him, I casually mentioned that I would like to someday teach in China. He immediately contacted his Chinese colleagues and before long I found a letter on my desk containing an invitation to come and spend a few weeks teaching at the Institute for Medical Sciences in Beijing. I thought about this for quite a long time and, after extensive negotiations with my department head, I finally received permission to answer their invitation. I prepared my teaching material and, at the end of May 1983, I boarded the Empress of Japan for a flight to Tokyo. Before leaving, I had contacted an organization that supported the underground church in China and arranged to obtain some Chinese bibles from them, which I hoped to smuggle in with me. For security reasons, I will leave to my reader's imagination how I got the bibles through customs and managed to deliver them to a member of the underground church.

The flight from Tokyo to Beijing is longer than expected as one has to make an extensive detour to avoid flying over North Korea. Upon landing in Beijing, I was met by my Chinese hosts who accompanied me from the airport along a beautiful tree-lined highway to the Beijing Hotel. My room was pleasant and every six hours they provided a big Thermos of hot water to accompany the large amount of complimentary tea. I also received a free copy of the China Daily newspaper in which the USA and the USSR were equally disparaged. It appeared that China's National People's Congress was having their annual meeting in the Great Hall at Tiananmen Square and much of the news was devoted to this. Many visitors from other countries seem to have been invited and every morning I could see the Toyota and Red Flag limousines arrive to take them from the hotel to the Congress. It may have been my imagination, but I had the impression that a few times at breakfast I was sitting right next to Baby Doc.

The lecture hall was on the top floor of the Institute and was quite large. It was packed with senior-looking people who, I was told, were teachers that had come from all over China to learn some immunology. I was expected to lecture for three hours straight, which was not easy given that the middle of the day in Beijing is the middle of the night in Winnipeg. I had a translator so after every sentence or two I had to pause so he could translate into Chinese. The only thing I can remember is that B cells and T cells are B *sheba* and T *sheba*. Before each session I invited them to join me in reciting a small prayer, which was my frail attempt at evangelization. My Chinese host was not too happy with this, but one of the students later told me that he had spent some time in Australia and the people he met there still wrote to him about Christianity.

The Catholic Church in China consists of an underground church loyal to the Holy Father and experiencing great persecution and an official overt church with "bishops" that are appointed by the government. I have been told that some of these "bishops" are secretly loyal to Rome but feign allegiance to the government in order to keep the official church alive. Pope Benedict XVI has asked recently for some attempt at reconciliation between the two groups. On Sunday, my Chinese host took me to one of the two functioning official Catholic churches in Beijing. There were a surprising number of young people at the service, which appeared to be pre-Vatican II, although it did not resemble anything that I had seen previously. Three men, who did not remove their Army caps, were also there and I assumed they came to keep an eye on things. There did not appear to be a Scripture reading or a homily, only lots of chanting by the older people.

After church my hosts took me for a ride in the country with a climb up Fragrant Hills where we stopped for a nice picnic. I noticed the countryside around Beijing was very beautiful. They also took me to a commune farm where they raise hundreds of Pekin ducks. There was a small infirmary containing a myriad of small drawers each filled with a different Chinese herb. In the main farmhouse I was shown a rather clever innovation for conserving scarce fuel supplies. The main bedroom had a small brick oven and the bed was built directly on top of the oven. In the winter, they lit a small fire in the oven which kept the bed warm, even though the remainder of the house was cold. On our return, we went to the zoo where I saw the famous panda bears. Then it was back to the hotel and the Institute for another week of lecturing. At the end of the week my hosts presented me with a Certificate of Appreciation and an exquisite silk tablecloth and several of the senior students thanked me with apparent sincerity. That evening I purchased a large box of chocolates before going to the dining hall and gave it to the young waiters and waitresses as a token of appreciation. They appeared very grateful and gave me royal treatment for the remainder of my stay.

On my last day in China my hosts arrived early to take me to the Great Wall. The farmers were harvesting wheat, which they spread on the roads to dry. Trucks and cars driving over the wheat helps to separate the chaff. It was quite slippery and our car slid all over the road. Although it was almost 38° C, climbing the Great Wall was exhilarating. Because it is so long, it is the only man-made structure that astronauts can recognize from outer space. In fact, the Chinese have a nice saying: "May our friendship last as long as the Great Wall."

On the way back from China I had a brief stopover in Tokyo, where I had the privilege of having lunch with Dr. Tomeo Tada, who was Professor of Immunology at Tokyo University and one of the top immunologists in the world. He passed away when he was quite young and about a year before he died he spent his time writing beautiful Japanese poetry. Later, as it was Sunday, I went to Mass at a Japanese Catholic Church and the priest there was very thoughtful. During his homily, he would say about five sentences in Japanese and then summarize what he said in English. How well I recall: "Where there is no love, put love." I left Japan about five o'clock Sunday evening and arrived in Vancouver at nine o'clock Sunday morning. I finally got back to Winnipeg at five o'clock Sunday evening, the exact same time that I left Tokyo. Thank God for a safe flight.

PART VIII

The Eternal Wedding Banquet

Gradually, as my knowledge of the faith was slowly increasing, I began to realize I needed to attend Mass and receive the Eucharist more often than just on Sunday. The Mass, that great dialogue of love between Christ and His heavenly Father and between Christ and His bride the Church. The Eucharist, where Christ's sacrifice on Calvary is rendered sacramentally present so that we might take part in it. The Mass, where history reaches its omega point and Christ comes to celebrate the eternal wedding feast with His bride even here and now.

There was a fairly well-known Catholic church in the center of Winnipeg and I began to ride my bicycle over there for daily Mass and it was there that I met three of the most saintly men that I have ever known. The first was an elderly blind layman named Joe Auffret. Joe and I became good friends and I learned a lot about the faith and about life from him. He loved to listen to Vatican Radio on his shortwave and to study the lives of the cardinals, although I am still uncertain how he did this inasmuch as he was blind and, therefore, could not read. I used to express my respect by referring to him as Joseph Cardinal Auffret.

The second two were Fr. Gerald Durocher and Brother Steven Capustinski. Everyone who knew Brother Steve was convinced that he was a walking saint. It was Fr. Durocher who recommended that I read *Faith and Certitude* by Fr. Thomas Dubay. In it Fr. Dubay presents a series of converging evidences for the existence of God that I found to be completely sensible and compelling. For those who might think that faith is anachronistic, Fr. Dubay points out that our whole intellectual life is based on faith, as one cannot prove that reason is reasonable. It is also impossible to be a complete skeptic, as this would entail saying it is true that there is nothing true. It is even impossible to be a complete subjectivist, as this would require a belief that it is a hard objective truth that all truth is subjective.

After about ten years of service at the university I was eligible for a short sabbatical leave. I decided that I should like to apply my knowledge of immunology to work on some problem affecting the poor in the developing countries. I had arranged to go and work on trypanosomiasis at the International Laboratory for Research in Animal Diseases in Nairobi, Kenya. Before I left, however, the head of Medical Microbiology convinced me to study schistosomiasis instead, because I knew a great deal about the IgE antibody response, which is a major mediator of protection against the disease. So I made alternative arrangements to go and work in the laboratory of Dr. Anthony Butterworth at Cambridge University in England.

PART IX

Court Fool of the King of Paradise

Because of my great love for the Eternal City, I decided to go and spend a few more days in Roma on my way to England. The Redemptorist Order was having their general chapter in Rome and, inasmuch as Czechoslovakia was still behind the Iron Curtain, my friend, Fr. Stan Liska, a Canadian-Czech priest, had to go and represent the Czech Church. I made arrangements to meet him, as we were going to be in Roma at the same time. I was staying in a pretty run-down hotel, but some German nuns had a health clinic close to St. Mary Major and they had a few guest rooms for visiting priests. How Fr. Liska managed to convince them to offer me one of these guest rooms remains a *magnum mysterium*. It was like a first-class hotel room and actually had hot water, all for $18 a night.

While in Rome, I took a day trip to Assisi, the home of St. Francis of Assisi, where I was able to pray at Poverello's tomb. Assisi sits near the top of a mountain and, inasmuch as it was never destroyed by war, it remains much as it was in St. Francis' day. I have read that only St. Paul followed Jesus with greater fidelity than St. Francis. Many of his contemporaries wanted to join him and this led to the foundation of the Franciscan order of friars. A few years before he went to his reward, St. Francis had a type of mystical vision and when it was over he bore in his body the five wounds of Jesus. There were some who considered St. Francis to be a fool, but as Chesterton so aptly put it, he was the court fool of the King of paradise. It has been said that the death of St. Francis was the stopping of that great heart that did not break until it held the world.

When I left Roma, I took the train to Calais, France, and then crossed the English Channel on a ferry from whose deck I could see the White Cliffs of Dover. Shortly after arriving at the London train station I purchased a ticket for a ride to Cambridge. The ticket master told me that the train would be leaving in a few minutes so I ran along the platform looking for an empty car. It was evening rush hour so every car I passed was full of passengers serried like sardines. When I reached the last car a man looked out and, given that my suitcase was full of books and weighed a ton, I asked him to help me. He reached down and swung my bag into the car while I jumped in after it only to find that I was in the baggage car. This was likely a blessing in disguise, as it was relatively uncrowded and one of my fellow baggage passengers recommended that I seek out Fisher House when I got to Cambridge.

PART X

Cercs and Somules

As it was the beginning of November, when I arrived in Cambridge it was already dark out. Aside from the magnificent colleges, a great deal of Cambridge consists of old English row houses which have acquired, with age, a rather fuliginous patina. However, when I arrived at 49 Maids Causeway, where I had rented a small third-floor apartment, I discovered that the interior of the homes was modern and quite cheerful. The house where I was staying belonged to a retired and rather renowned professor of German literature. He had converted one of the rooms on the first floor into an office where he had a beautiful antique desk. Because he was an expert in the German language, he was recruited, during the Second World War, to decode German messages and he knew the Canadian spymaster, code name Intrepid, who was the head of British intelligence at the time. Once when he was saying grace for us he also gave me a demonstration of his proficiency in Latin.

The next morning I took my bicycle and rode off to the Pathology Department to meet with Dr. Anthony Butterworth and the rest of his laboratory, to begin what would be amongst the happiest four months of my life. As mentioned, I had come to study schistosomiasis and the first order of business was to learn how to maintain the life cycle. Whereas the vector for most parasitic diseases is a biting insect, the vector for schisto is a particular type of snail. Because of this, it is especially endemic in the Nile River Valley and in other places where people work in rice paddies. The infected snails shed cercariae which penetrate the skin of humans and develop into schistosomulae. The latter travel to the liver where males and females mate and then the females produce eggs which are excreted back into the water system. Here the eggs infect new snails and the cycle is complete. Ultimately, I hoped to induce high-level IgE responses to cercariae and schistosomulae in mice to determine whether this would confer protection against challenge and then to try to identify the antigens responsible for protection.

In addition to Butterworth, I also met and worked with Dr. David Dunne, who was a senior scientist in the lab and a real expert on immune responses to schistosome infections. Whatever I learned during my stay was a result of his clear thinking and generous assistance. The only request of his that I did not obey, and now wish that I had, on occasion, was to join him after work for a few pints in one of Cambridge's numerous pubs.

Shortly after my arrival, I decided to look up Fisher House, the Catholic chaplaincy in Cambridge. It was named after St. John Fisher, who was a friend of St. Thomas More and the only English bishop to refuse to agree to Henry VIII's declaration that he, and not the pope, was now the supreme head of the church in England. All this, of course, was to enable Henry to divorce Catherine of Aragon and commit adultery with Anne Boleyn. Ultimately both Fisher and More were martyred at Tower Hill for their fidelity to God and to the truth. Just before he died Fisher exclaimed: "Send the king good counsel," while More declared: "I die the king's good servant, but God's first."

The main chaplain at Fisher House was a Benedictine monk named Fr. Chris Jenkins. I began to go to the chapel for morning Mass and the Liturgy of the Hours and, because Fisher House was very close to the Pathology Department, I often also went in the evening for Vespers. I must admit that I was deeply impressed by the sincerity and piety of the young British students who also came there to pray. St. John Vianney said that there is nothing easier than praying and nothing more comforting.

Once a year, the chaplaincy sponsored a fairly prestigious lecture series and, while I was there, these were given by Professor Keith Ward from King's College London. The titles of his talks were "What scientists are saying about God" and "What philosophers are saying about God." This was my first encounter with philosophy which subtends most of Catholic theology. Both science and philosophy are a search for the truth, although their methods are somewhat different. Whereas science uses inductive reasoning about a series of observable facts to come to a general conclusion, philosophy starts with some self-evident first principle or axiom and uses deductive reasoning to come to other general conclusions. For example if one starts with the premise that

God is perfect, one can deductively conclude that God is
One, as you cannot have two perfect beings that differ from
each other. Ultimately, philosophy tries to answer questions
such as What does it mean to be, What does it mean to
know, Can anything really be known given that things are
always changing, What is truth?

Things were progressing quite well in the lab, when, on
Christmas Eve, some disastrous news arrived from Winnipeg. Prior to my leaving on sabbatical, the head of Immunology had recruited a fairly eminent allergist from Belgium
and was grooming him to become the new director of our
Medical Research Council Group for Allergy Research. I
felt that the renewal of the Group grant and, hence, the
renewal of my position, absolutely depended on him.
However, it seems that he had a major disagreement with
the department chief and current director and, without too
much warning, took his entire lab to Montreal. I had to
wait to deal with this when I returned to Winnipeg.

While in Cambridge I was vaguely aware that there was a
place called Blackfriars. On the last Sunday before I was to
return to Winnipeg, I rode my bicycle over to investigate
and consequently had my first and only encounter with
the famous English Dominicans. I stayed for Mass and the
impressive sermon gave me a good indication of why they
are also known as the Order of Friars Preachers. I spoke
with one of the friars after Mass about the needs of the
Church in China and I can still hear him tell me as I was
about to leave: "Jolly good to have met you, Robert."

It was sunny and warm when I left Cambridge and, being
the beginning of March, it was cold with a great deal of

snow when I got back to Winnipeg. I was quite disheart-
ened because I felt that the Group grant likely would not
be renewed. To make matters worse, newly promulgated
Medical Research Council regulations for Group grants
prohibited principal investigators from seeking, during
the tenure of the grant, a more permanent position at an
alternative institution. Consequently, even though I did
not have another job to go to, I decided it was time for
me to leave. As it turned out the Group grant was given a
provisional renewal for one year and then was terminated.

It took me several unhappy months to find another job,
during which time, I kept busy sending resumes and
reading immunology literature. One of my old friends
at the Medical Research Council suggested I contact Dr.
Brian Barber in the Department of Immunology at the
University of Toronto, as he had just obtained a large five-
year grant from the Ontario government to promote collaboration between government, university, and industry. He
agreed to hire me as a research associate and so it was back
to Toronto to start all over again. I was not very optimistic
about the technology that he was trying to develop but I
had to accept what was offered to me.

I spent the next four years in Toronto working as hard as I
could. Not too much came out of it except that I learned
to work with virus and to carry out assays for cytotoxic T
lymphocytes (CTL). My private study of Catholic theology
was also reaching a more advanced level as I was beginning
to read works by the brilliant Joseph Ratzinger and some
of the great monastic writers. There are those who say that
monasteries are a prolepsis of the New Jerusalem. All of
this was greatly facilitated by the fact that the Daughters of

St. Paul had a large Catholic bookstore in Toronto. While in Toronto I also met the Jesuit Fr. John Perry, who was completing his doctorate in theology. He has since written a book entitled *Torture: Religious Ethics and National Security* (2005), which won the Pax Christi USA award for "Best Book in 2006." The last I heard from him he was at the Kofi Annan Institute for Conflict Transformation at the University of Liberia in Africa and, I might add, that this was during the Ebola outbreak, so I hope he is still alright.

It was also about this time that I began to ask myself what Jesus actually meant when He commanded that we love one another as He loved us. What does love, as opposed to being in love, really mean? After much thinking I came up with three definitions which seemed to satisfy me. First, it meant to seek the good of another even at the cost of self-sacrifice. Second, it is what a mother feels for her child. Third, it is a manifestation of the Trinitarian nature and goodness of God in whose image we are made. The late Fr. Benedict Groeschel has said that our struggles with our own lack of love should keep us humble and compassionate toward others. Of course, all of this was subsequently made even clearer by Pope Benedict XVI's brilliant and lucid encyclical *Deus caritas est*.

I also began to wonder about the notion of space time. Light travels at the speed of 186,000 miles per second and a light year is how far light travels in one year. So if you look at a star that is 1,000 light years away you are seeing that star not as it is today but how and where it was 1,000 years ago. In other words, you are looking back in time. I still think a lot about what that actually means. This, as well as other amazing discoveries of modern cosmology, such as the notions that space is not nothing but can actually be bent, time is not absolute, the rate of expansion of the universe is increasing rather than decreasing, and that all of these phenomena are under the influence of a mysterious dark energy, have lead me to tentatively conclude that when we go far enough back in space or time, we do not know for certain what we are dealing with. Perhaps if I understood Einstein's theory of relativity it would become clearer. In any case, I am now quite skeptical when I hear people speak with such certainty about billions of years.

Near the end of the five-year technology fund, I decided to make one more attempt at trying to fulfill my original goal of using my knowledge of immunology to work on some problem affecting the poor in developing countries. I wrote to many investigators studying a variety of parasitic diseases, but money was in short supply and I had no offers. I had been corresponding with several Salesian priests in Central and South America and, after mentioning my dilemma, they promised to pray for me. Immediately afterwards, I came across a publication dealing with an attempt to develop a vaccine against malaria from the Walter Reed Army Institute of Research in Washington, DC. I straightaway wrote to them and within a few days the person in charge called me and asked: "How did you know that I have a position available? I have not even advertised it." After considerable discussion they decided to hire me as a contractor on a temporary basis. It turned out that I would be there for almost twenty years.

PART XI

Vaccine Studies

Malaria is probably the worst of the so-called great neglected diseases. At that time there were hundreds of millions of new cases every year with about one million children dying; the majority in Sub-Saharan Africa where one encounters the deadly falciparum version of the parasite. Whereas with schistosomiasis the vector is a snail, with malaria the vector is the Anopheles mosquito. When the mosquito bites it injects a small number of sporozoites into the blood stream which quickly travel to the liver to invade hepatocytes. For the next twelve days the infected individual remains asymptomatic, while the sporozoites multiply into thousands of merozoites. The latter then burst from the liver and rapidly invade red blood cells, where they further multiply into many new merozoites. When a normal mosquito ingests infected blood from a carrier it, in turn, becomes infected and the life cycle is complete.

In the 1980s, prior to my arrival at Walter Reed, investigators had prepared monoclonal antibodies against both mouse and human malaria sporozoites and using these in combination with gene cloning (which I described previously) they discovered that the sporozoite is coated with a densely packed layer of a single protein, appropriately referred to as the circumsporozoite protein (CSP). The Plasmodium falciparum CSP exhibited an unusual structure in that it had a large central repeat region that consisted of over forty tandem copies of a block of four amino acids: asparagine-alanine-asparagine-proline, which in biochemistry shorthand is designated NANP, as well as a sizeable flanking region on either side of the repeat. It turned out that this repeat region was a primary target of antibody responses and that antibodies against NANP could confer protection against sporozoite challenge.

Now the United States Army had a vested interest in making a vaccine against malaria for its troops. It recalled how, during the Second World War, General MacArthur had said that in order to invade the Philippines he would need three armies: one to do the fighting, one to be sick in the hospital with malaria, and one to be recuperating from malaria. They had learned the same lesson in Vietnam. I should add that many Army doctors had seen children die from malaria so there was also an altruistic motive for wanting a vaccine. It was known that hepatitis B virus surface antigen (HBsAg) was safe and immunogenic so the Army entered into a collaboration with GlaxoSmithKline Belgium to use gene cloning technology to produce a hybrid HBsAg-CSP protein, known as RTS,S. It is well-known in immunology that materials referred to as adjuvants will potentiate an immune response, so to make RTS,S even more immunogenic, it was mixed in an oil and water emulsion, with two powerful adjuvants known as MPL and QS-21. The whole trial vaccine was referred to as RTS,S/AS02A.

Before allowing RTS,S/AS02A to go into humans, the Food and Drug Administration wanted to first ascertain whether it was immunogenic in mice and this is when I arrived at Walter Reed. My first task was to demonstrate strong immunogenicity in mice and, in collaboration with others in the lab, this was quickly accomplished. So the candidate vaccine was ready for phase I (safety and immunogenicity) and phase IIa (small-scale safety, immunogenicity, and efficacy) trials in humans.

Because blood-stage malaria can be eliminated by drug treatment, it is feasible and ethical to carry out a challenge trial. Volunteers are immunized with a candidate vaccine and then they, as well as a non-immunized control group, are challenged by infectious mosquito bites. Non-immune controls should develop malaria beginning around twelve days post-challenge, while immunized volunteers, given an ideal vaccine, should remain completely protected. Control and any immunized subjects who develop malaria are treated immediately with anti-malaria drugs. This type of vaccine trial is very expensive as it requires infected Anopheles mosquitoes which, in turn, entails the large-scale maintenance of the entire malaria life cycle in a highly controlled insectary.

The initial trial involved 8 malaria naïve American volunteers, who received 3 doses of the candidate RTS,S/AS02A vaccine at time zero, 1 month and 7 months and then 2 weeks later they, along with unimmunized controls, were challenged with bites from 5 infected mosquitoes. As was expected, between 12 and 14 days post-challenge all of the controls came down with malaria. How well I remember however, even though this happened over 20 years ago, the mounting excitement as day after day Drs. Bruce Weldy and Ted Hall examined blood smears from the immunized group and found that, even 2 months later, 6 out of 8 of the volunteers were still malaria free. This was the first time in history that anyone had ever induced protection against malaria using a sub-unit protein vaccine.

Over the next several years many more vaccine trials were conducted at Walter Reed to optimize the number of doses of the vaccine, the immunization regimen, and the adjuvant and to try to identify the immune response(s) that mediated protection. My own contribution was to determine whether cell mediated immunity was contributing to protection and to characterize the nature of the protective responses. Over the course of these many trials the vaccine was repeatedly shown to confer about 50% protection in naïve adult volunteers, when administered in a 0, 1-month, 2-month regimen. The fact that 50% of the volunteers were protected and 50% were not, allowed a comparison of the cell mediated responses in protected versus non-protected groups, which greatly facilitated identification of the responses that appeared to be mediating protection. A new adjuvant, which was designated AS01B and which consisted of MPL and QS21 in a liposome formulation, instead of an oil and water emulsion, was also tested and found to be superior to AS02A. Once vaccine antigen dose, adjuvant, and immunization regimen were decided, the number of volunteers tested was increased to over 100 so as to obtain sufficient data to enable hard-core statistical analyses. More about this later.

After I arrived in Washington, I decided that it would be wise for me to have a spiritual director. The Dominicans have a very large House of Studies in Washington located next to the Catholic University of America and, given that I had come to admire the theological traditions of the Order, I decided to enquire whether anyone there would be willing to act as my director. This is how I got to meet the very wise and kind Fr. Pierre Conway, O.P., whose biological father, interestingly, was also from Winnipeg. Over many years he was always willing to see me and to offer his sympathetic and prudent advice. He also, I should add, became my confessor and after suitable admonition shrived me of my many sins. I never realized exactly how famous a scholar he was until after he passed away and there was a large article about him in the Catholic paper. It seems that he even translated some of St. Thomas Aquinas' works from Latin into English. After he passed on, he donated his body to science, and people who knew him said that's just like Fr. Pierre, giving to the very end.

About halfway through my tenure at Walter Reed I was asked, once again, to attend the International Congress of Immunology, this time in Stockholm. As in the past, I was able to meet many immunologists including Brian Barber, my former colleague from Toronto. One of the highlights of this Congress was that it included a fair number of debates between eminent immunologists, who had to defend their divergent opinions about a particular immunological theory.

PART XII

A Shower of Roses

After the Congress was over, I decided, as I had previously, to spend a week of my summer vacation in France. Upon arriving in Paris, I took the train to the small town of Lisieux to visit the Carmelite convent and reliquary of St. Therese of the Child Jesus, one of the most famous saints of the Church. It seems that there are those who become saints by living lives of heroic virtue and then there are those, like St. Therese, who become saints because they were called by God, at an early age, to accomplish some specific mission for the benefit of the Church. St. Therese lived during the second half of the nineteenth century and entered the Carmel of Lisieux when she was only fifteen. Almost immediately she experienced the dark night of the soul, which is a sharing in Christ's feeling of abandonment on the Cross. She often said that God was leading her by a subterranean way where there is no light, only darkness.

While in Carmel, Therese wrote a fair number of poems, including:

> Their loss is gain who all forsake
> To find your love, O Jesus mine!
> For you my ointment jar I break,
> The perfume of my life is thine.

Near the end of her life on earth, at age twenty-four, the mother superior of her convent, who was also Therese's blood sister, commanded her to write an account of her life. In obedience, Therese composed her autobiography which she called simply *The Story of a Soul*, and it was here that she described her little way of spiritual childhood. About this time, she also developed tuberculosis, which spread throughout her body and caused her great suffering. Prior to her death, she told her Carmelite sisters that they will not have time to miss her; the postman will keep them very busy on her account. In prophesying the number of miracles that would be accomplished at her intercession, she said: "You will see, it will be like a shower of roses."

It is the custom of Carmel, on the death of one of the community, to publish a short account of the sister's life and to send it to other Carmelite convents. In the case of St. Therese, it was decided that *The Story of a Soul* would serve the purpose. The book experienced an immediate and enthusiastic reception. From convent to convent it was read with alacrity, and from the convents it was lent to others. By 1932 the circulation had risen to over 2,000,000 copies and it has since been translated into 35 languages. Upon reading the book people began to ask Therese for her intercession and thus began the promised shower of roses. From every corner of the world there came the report of miracles, such that by the 1930s over 1,000 letters a day were arriving at the Carmel of Lisieux. Obviously, this holy and brave daughter of Normandy was fulfilling her pledge to keep her sisters too busy to miss her.

I returned to Paris and the next day boarded the TGV, a high-speed train, for a ride to the Pyrenees and Lourdes, the village of St. Bernadette. It was here, at the famous grotto, that the young Bernadette Soubirous had several visits from Our Lady. When Bernadette asked the Blessed Mother's name, she replied: "I am the Immaculate Conception." I have no definitive teaching about this but I imagine that Our Lady revealed herself this way to confirm Pope Pius IX's dogmatic definition that Mary, the mother of Jesus, was conceived immaculate, that is, Mary's soul was preserved from original sin and from its effects from the moment of her conception. One can think of at least two reasons why the Immaculate Conception was necessary. First, as the Second Person of the Blessed Trinity, Jesus is always a divine Person, but He is a divine Person who assumed a human nature. Jesus drew His human nature from His mother. Because He is both God and man, Jesus could not have original sin or even a propensity to sin in His human nature, as it was in hypostasis with His divine nature. Second, someone on behalf of the entire human race had to be able to give an unlimited yes to God's unlimited Word. If Mary's yes to God (her fiat) had contained even a hint of a demurral, of a "this far but no farther," a stain would have clung to her faith and the Child could not have taken hold of the entirety of human nature. Her will had to be entirely in accord with the will of God.

One of the most memorable events for visitors to Lourdes is the evening candlelight procession. Being one of 30,000 pilgrims, each with a lit candle, processing through the darkness is something I won't soon forget. Eventually, Bernadette left Lourdes to enter the Ursuline convent in Nevers where she remained until her death in 1879. She was canonized in 1933 and her incorrupt body can be seen today in a crystal reliquary in the convent's chapel.

About this juncture of my stay in Washington, Pope John Paul II passed away. In my opinion this one man alone kept the entire second half of the twentieth century from going to perdition in a basket. The secular media likes to honor him for his contribution to the downfall of communism in Eastern Europe, but he did so much more for the Church and for mankind. It is not my purpose here to write his biography; one can refer to George Weigel's *Witness to Hope* and its sequel *The End and the Beginning* to read about this remarkable life. I would like to mention, however, that as he lay dying those nearby heard him say: "I have sought you and you have come for me and I thank you." At first those who heard him thought he was talking about the large crowd of mourners that had gathered in St. Peter's Square to pray for him. Later they came to believe that, as death neared and the veil began to lift, he was catching sight of the angels or maybe even Our Lady coming to escort his noble soul into the presence of God.

PART XIII

The Great GKC

My reading of Catholic literature continued with ever greater precision and intensity. It was at this point that I encountered, for the first time, the writings of the great G. K. Chesterton, who lived and worked in England until his death in 1936. He was not a renowned Catholic philosopher like Jacques Maritain or Etienne Gilson or Bernard Lonergan, nor a great Catholic theologian like Reginald Garrigou-Lagrange or Hans Urs von Balthasar. He was, however, someone who loved paradox and humor and poetry and excelled at all three in his prolific writing. The renowned atheist, George Bernard Shaw, affirmed that Chesterton was a man of colossal genius. There are literally hundreds of famous quotes from his creative pen and I would like to share but a few of them:

> "Love wants to take you to the altar, but an altar is a place for sacrifice."

"All science, even the divine science, is a sublime detective story. Only it is not set to detect why a man is dead, but the darker secret of why he is alive."

"Those who attack the faith in the name of reason usually end up attacking reason instead."

"Christianity has appeared to die many times but it always rose up again because it has a founder who knows the way out of the grave."

"There are those who insist that a throne is only a chair. I would rather believe that every chair is a throne."

In a poem to honor Bethlehem, he referred to it as:

"The end of the way of the wandering star,
The place where things cannot be, but are,
The place where God was homeless
And all people are at home."

Pope Pius XI made Chesterton a Knight Commander of the Order of St. Gregory. The cause for his beatification has now been opened.

I also started to read the shorter works of the brilliant and ethereal Hans Urs von Balthasar. He claims that the first thing that the Cross does is to cross out the world's word and replace it with a wholly-other Word that the world does not want to hear. This Word is God's language and it is so stupendous that it can never be considered past tense. He also discusses something that I had begun to think deeply about: while Mohammed never thought he was Allah, and

Moses never considered that he was Jehovah, Jesus claimed that He is God. He called God His Father and said: "I and the Father are one." If you think about this it should make you feel like the ground has been taken out from under you and that your head has been plunged into instant vertigo. If Jesus believed this and it wasn't true then He could be considered a madman. If He did not believe it but still said it then He would be a liar and a blasphemer. But this is where it becomes interesting. No one would ever think Jesus to be mad or a blasphemer because He is so wise and so good. That leaves only the third option: that He believed it because it is true. Von Balthasar maintains that we must come to believe that there is such a thing as Absolute Love and that there is nothing higher or greater than it. This Love will be God's last Word to the world before heaven and earth pass away. Pope John Paul II elevated von Balthasar to the College of Cardinals but he passed away just as he was leaving to go to Rome to receive his red hat.

PART XIV

A Vaccine at Last

The RTS,S/AS01B vaccine has now undergone large scale phase III clinical trials involving thousands of children at eleven different malaria endemic sites in Africa. This was very expensive as in most places the infrastructure had to be built up from zero and locals had to be hired and trained. Overall, for children over five months of age, the vaccine still confers 50% protection for one year. The World Health Organization had hoped for 80% protection but, because malaria is such a serious problem, they will likely license RTS,S/AS01B. In the meantime, two new publications have appeared in the literature demonstrating that either varying the immunization regimen as well as the amount of the third dose, or combining RTS,S/AS01B with an adeno-virus construct containing non-CSP malaria antigens, can both raise the degree of protection up to the required 80%. If this level of resistance holds up in the field, the RTS,S/AS01B vaccine, in combination with pesticide treated bed nets, may succeed in eradicating falciparum malaria from the face of the earth. I suspect that this will merit a Nobel Prize for the original inventors of the vaccine.

PART XV

A National Tragedy

While it is a blessing that this great collaborative effort, involving many years and millions of dollars, may end up saving the lives of countless millions of children, it is, at the same time, very disturbing to me that millions of other children are dying unnecessarily every year in the developed countries of the United States and Canada. It all began during the early 1970s, while I was still in Winnipeg, when something happened that badly shook my faith in democracy. In the United States it was the infamous *Roe v. Wade* decision that, not by the will of the people, but by edict of the Supreme Court, legalized abortion during the first six months of pregnancy. The justices maintained that they found a constitutional right to abortion in the penumbra (shadow) surrounding the constitutional right to privacy guaranteed by the Fourteenth Amendment. This, of course, is ludicrous as one can find whatever one wants in the so-called penumbra. Even the very strongly pro-abortion Justice Ruth Bader Ginsburg has acknowledged that *Roe v. Wade* was wrongfully decided. In addition, Norma

McCorvey, the "Jane Roe" of *Roe v. Wade*, has since admitted that the whole thing was a lie; that she had not, as was claimed, been raped. She later converted to Catholicism and worked hard to have the decision overturned and her name cleared. At present about sixty million, I repeat, sixty million babies have been condemned to a premature death in the United States as a result of this tragedy. I can only ask what kind of a country not only tolerates but actually enshrines in law and celebrates as a right the destruction of sixty million of her own children. The situation in Canada was not much better; the former Prime Minister Pierre Trudeau buried the legalization of abortion in a large omnibus bill that he rammed through Parliament and then a special committee in the Canadian Senate, that so-called chamber of sober second thought, rubber-stamped it in less than a day.

If that wasn't enough, the media in both countries, who are supposed to tell the unbiased truth, did their best to convince the public to support abortion on demand. With conviction unencumbered by knowledge, they used all of their usual subterfuges such as constantly speaking about women's health when they knew full well that in Roe's companion case, *Doe v. Bolton*, the court defined health to mean whatever you wanted it to mean. Nevertheless, the media constantly made it sound as if abortion was performed only to prevent some direct adverse effect that a continuation of the pregnancy would have on a woman's physical health. Similarly, what everyone knew to be a baby, suddenly became a worthless fetus, taking for granted that the general public was too benighted to realize that *fetus* is Latin for baby. There are many other egregious examples of bias that I could write about here as well. Of course, once the press got abortion on demand at any time and for any reason, they did not want to rock the boat so they made it a non-issue. In the Soviet Union you had censorship *of* the press while in the United States and Canada you had censorship *by* the press.

It was also disturbing that the largest taxpayer-funded abortion provider was Planned Parenthood. This organization was started by Margaret Sanger who maintained that American blacks and other "inferior classes" should all be sterilized and one wonders why many of Planned Parenthood's abortion mills are now located in black neighborhoods.

One of the solutions to the abortion problem is given to us by Pope St. John Paul II in his encyclical *Evangelium vitae* (Gospel of life): one must be radically for the child and radically for the woman. Provide her with all that she needs so that she will want to choose life. It seems that people don't reach sanity until they reach sanctity.

PART XVI

The Doctrine of Aseity

I have retired now but I continue trying to keep up with the immunology literature. I am also studying many of the writings of Pope Benedict XVI, whom many consider to be the most intelligent man alive today and, as described above, of the late Hans Urs von Balthasar whom many believe to be the greatest theologian of the twentieth century. I have not yet developed the courage to tackle the *Summa Theologica*; I am not yet and probably never will be a Thomist.

All of my reading has brought me to think about what I now consider to be the most important question one can ask. If nothing, by which I mean the absolute absence of anything and everything, cannot on its own evolve into something, why then is there something? The only answer is the doctrine of aseity which proposes that there must be a Being that holds within itself the cause for its own existence; a Being who is self-existent by nature; a Being for which "that it is (existence)" and "what it is (essence)" are completely one. If there is such an absolute and necessary

Being and, if this Being is sufficient unto itself, it is almost even more mysterious why there should exist something else. Von Balthasar maintains that only a philosophy of love can account for our existence, though not unless it also interprets the essence of finite being in terms of love. I also wonder about the tremendous order that exists in the universe and ask how it came about given that the law of entropy states that, left on their own, things always tend to disorder. But that can be grist for another mill.

I conclude this reflection on a little life with a stanza from the poem "God's Grandeur," by the renowned Jesuit poet Gerard Manley Hopkins:

And for all this, nature is never spent;
There lives the dearest freshness deep down things;
And though the last lights off the black West went
Oh, morning, at the brown brink eastward, springs—
Because the Holy Ghost over the bent
World broods with warm breast and with ah! bright wings.

Magnificat anima mea Dominum

About Leonine Publishers

Leonine Publishers LLC makes fine Catholic literature available to Catholics throughout the English-speaking world. Leonine Publishers offers an innovative "hybrid" approach to book publication that helps authors as well as readers. Please visit our web site at www.leoninepublishers.com to learn more about us. Browse our online bookstore to find more solid Catholic titles to uplift, challenge, and inspire.

Our patron and namesake is Pope Leo XIII, a prudent, yet uncompromising pope during the stormy years at the close of the 19th century. Please join us as we ask his intercession for our family of readers and authors.

Do you have a book inside you? Visit our web site today. Leonine Publishers accepts manuscripts from Catholic authors like you. If your book is selected for publication, you will have an active part in the production process. This book is an example of our growing selection of literature for the busy Catholic reader of the 21st century.

www.leoninepublishers.com

www.ingramcontent.com/pod-product-compliance
Lightning Source LLC
Chambersburg PA
CBHW071429040426
42445CB00012BA/1314